Túlio Silva Lara
Amauri A. Alvarenga

Physiological changes in tomato seeds induced by osmopriming

Túlio Silva Lara
Amauri A. Alvarenga

Physiological changes in tomato seeds induced by osmopriming

A strategy to increase seed vigor and initial seedling growth

ScienciaScripts

Imprint

Cover image: www.ingimage.com

This book is a translation from the original published under ISBN 978-613-9-67946-1.

Publisher:
Sciencia Scripts
is a trademark of
Dodo Books Indian Ocean Ltd. and OmniScriptum S.R.L publishing group

120 High Road, East Finchley, London, N2 9ED, United Kingdom
Str. Armeneasca 28/1, office 1, Chisinau MD-2012, Republic of Moldova, Europe
Printed at: see last page
ISBN: 978-620-8-18776-7

SUMMARY

ACKNOWLEDGMENTS

To God who has never abandoned me, even in the most difficult times.

To my parents, who gave up many things for me to realize my dreams, as well as teaching me character and honesty.

To my brother, who will always be by my side.

To my nephew, who brought happiness back home.

To my girlfriend and friend Thais, who has always been by my side, listening and supporting me.

To my uncles, especially Uncle Rogério, for his conviction that you can only make your dreams come true if you work hard.

My aunts, especially Wanir and Elizete, for the hours of relaxation in Guapé.

To my cousins, especially Nathalia and Taciane, who have become sisters.

To my grandparents up there, who are cheering our victory.

My grandmothers for their example of faith.

To my friends from Guapé, childhood friends, friends for life.

To my friends at Rep. Balaio and Rep. Lobo Mau, who have become great friends and brothers.

To my friends from home, for their companionship.

To Mrs. Hilda, who has become a very special person.

To Professor Amauri Alves de Alvarenga, for guiding me since the beginning of my studies.

To Professors Luiz Edson and Donizeti for trusting me to use their laboratories.

To my friends in Plant Physiology.

To my friends in the laboratory (Sara, Amanda, Jean, Mariana, Lara, Tàssia, Fernanda) without whom none of this would be possible.

To everyone who supported me in this achievement, making it everyone's achievement.

To the Federal University of Lavras (UFLA) and the Plant Physiology Program for the opportunity.

To the Minas Gerais State Research Foundation (FAPEMIG) for the grant.

SUMMARY

Osmotic conditioning is a technique that can be used to improve the germination performance of seeds. The aim was to identify the influence of conditioning agents on the germination metabolism and physiological performance of tomato seeds. Tomato seeds (*Lycopersicon esculentum* Mill.) cv Santa Clara were conditioned in solutions of polyethylene glycol 6000 (PEG) and potassium nitrate (KNO_3) for 2, 4 and 6 days. The osmotic agent solutions (v/v) were: 100% PEG for 2, 4 and 6 days; 100% KNO_3 for 2, 4 and 6 days; and combinations of 50% PEG + 50% KNO_3 for 2, 4 and 6 days, as well as the control (unconditioned seeds). Two trials were carried out, the first analyzing the activity of some enzymes of the antioxidant system and the activity of nitrate reductase and their possible contributions to germination. The second trial analyzed the concentration of some carbohydrates and their possible contribution to germination and initial seedling development. In the first trial, osmotic conditioning with 100% KNO_3 for 4 days and 6 days resulted in higher nitrate reductase and superoxide dismutase activity. In addition, the speed of emergence and seedling formation were higher. In the second trial, a higher germination rate was observed under ideal conditions of water stress and low temperature in the conditioned seeds, regardless of the incubation medium, compared to the control. Lower total soluble sugar concentrations were also observed in conditioned seeds. Conditioning in 100% KNO_3 for 4 and 6 days showed the highest germination, normal seedling formation and emergence rates, as well as the highest nitrate reductase and superoxide dismutase activity. In addition, osmotic conditioning, regardless of the solute and the incubation period, increased the germination speed index under ideal, water-stressed and low-temperature conditions, and there was a decrease in the concentration of total soluble sugar in conditioned tomato seeds.

Keywords: Conditioning. Nitrate reductase. Antioxidant system.

CHAPTER 1 GENERAL INTRODUCTION

1 INTRODUCTION

World production of tomatoes (*Lycopersicon esculentum* Mill.) in 2009 was around 141 million tons from approximately 5 million hectares (FOOD AND AGRICULTURAL ORGANIZATION - FAO, 2010). Brazil is the world's ninth largest tomato producer, with an estimated production of approximately 4 million tons on an area of approximately 65,391 hectares (INSTITUTO BRASILEIRO DE GEOGRAFIA E ESTATISTICA - IBGE, 2011). One of the main reasons for such expressive figures is the increase in consumption growth: between 1985 and 2005, the world's *per capita* demand for tomatoes grew by around 36% (CARVALHO; PAGLIUCA, 2007).

Alongside the growth in production, the tomato seed market is also expanding, accounting for around 27% of all vegetable seeds sold in Brazil and generating approximately 12 million reais a year (ASSOCIAÇÃO BRASILEIRA DO COMÉRCIO DE SEMENTES E MUDAS - ABCSEM, 2009). In addition, the demands and high competitiveness of the market for good quality and differentiated products encourage companies to look for new technologies in order to produce quality seeds. Based on the assumption that seed quality is the starting point for crop development and productivity.

There are techniques used to improve seed performance in the field, as well as to improve germination under adverse temperature and humidity conditions (VARIER; VARI; DADLANI, 2010). Among the most widely used today is osmotic conditioning (*seedpriming, osmoconditioning, osmopriming, osmoticpriming*). According to Bradford (1986) and Heydecker and Gibbins (1978), the technique is based on immersing the seeds in a solution with a low osmotic potential, in which the absorption of water by the seed is limited, so that the germination process is not completed, only initiated.

Many solutions can be used in the osmotic conditioning technique. Saline solutions allow for better aeration of the solution and it is easily removed from the seeds during washing; the ions present in this solution are absorbed by the seed. Inert solutions have the advantage of not penetrating the seeds, but do not allow for uniform aeration of the solution and are difficult to remove after treatment (NASCIMENTO, 2004).

The main benefits provided by this technique are increases in germination and emergence rates, as well as increasing the percentage of germination, improving seedling vigor and breaking dormancy (VARIER; VARI; DADLANI, 2010). The benefits come from improvements that occur at cellular level: induction of nuclear DNA replication, accumulation of β-tubulin and induction of assembly of the microtubular cytoskeleton network (CASTRO et al., 2000; LANTERI et al., 2008). It also provides a significant improvement in the antioxidant system and decreases the concentration of malondialdehyde (MDA) (CHEN; ARORA, 2011; EL- ARABY; HEGAZI, 2004; KIBINZA et al., 2011). There is also an increased expression of genes encoding LEAs proteins, *heat shock*, enzymes involved in DNA replication, as well as inhibiting the action of proteases (CORTEZ-BAHEZA et al., 2007) and causing changes in reserve metabolism and cellular homeostasis (FAROOQ et al., 2010). The aim was to determine the influence of osmotic agents on the germination metabolism and physiological performance of conditioned tomato seeds.

2 . THEORETICAL FRAMEWORK

2.1 Osmotic conditioning

The osmotic conditioning technique has been used to increase the speed of germination, reducing the time between sowing and the last protrusion, as well as increasing the tolerance of seeds to unfavorable environmental conditions during sowing. This technique can be summarized as controlled absorption of water by the seeds, initiating the metabolic processes for germination, but without protrusion. The osmotic potential of the solution in which the seeds are immersed controls the entry of water into the seed (BRADFORD, 1986; HEYDECKER; GIBBINS, 1978).

Conditioning acts on phases I and II of germination, allowing the seed to remain at a moisture content just below that ideal for root protrusion (POSSE et al., 2002). As a result, metabolic processes begin before planting, which gives the seed an advantage over unconditioned seeds. In addition, conditioning provides a slower and more controlled absorption of water than in the field, where there is a longer period for repairing cell structures (VARIER; VARI; DADLANI, 2010).

2.2 Different osmotic agents

Different solutes can be used in osmotic conditioning, such as polyethylene glycol (PEG), potassium nitrate (KNO_3), potassium phosphate (K_3PO_4), monobasic potassium phosphate (KH_2PO_4), magnesium sulphate ($MgSO_4$), manganese sulphate ($MnSO_4$), magnesium chloride ($MgCl_2$), sodium chloride (NaCl), sodium nitrate ($NaNO_3$), mannitol, glycerol, among others. All these compounds are used to adjust the osmotic potential of the solution (NASCIMENTO, 2004).

Saline compounds allow for better aeration of the solution and are easily removed from the seeds during washing. In some cases, ions from salts are absorbed by the seeds, negatively affecting germination; on the other hand, this absorption can activate metabolic pathways that would not naturally be active, which can favor germination (NAWAZ et al., 2012; PACE et al., 2012). Inert solutions, such as PEG, have the advantage of not penetrating the seeds because of their high molecular weight. However, PEG solutions do not allow for uniform aeration. In addition, PEG is difficult to remove after treatment, as well as being expensive (NASCIMENTO, 2004).

However, the most commonly used solutions are PEG and KNO3, and the differences in physiological responses will depend on the type of solution and the species. Chen and Arora (2011) and Kibinza et al. (2011) worked with spinach and sunflower seeds, respectively, in PEG solution. Arin et al. (2011) and Kissmann et al. (2011) worked with onion and jacaranda seeds, respectively, in a KNO solution$_3$. In tomato studies, this preference was maintained by Bocian and Holubowicz (2008), Govinden-Soulange and Levantard (2008) and Nawaz et al. (2012).

2.3 Effect of nitrate on germination metabolism

Nitrate is an important source of nitrogen for plants, but it is also a signaling molecule that controls various aspects of plant development, including seed dormancy (ALBORESI et al., 2005). Nitrogen from the soil is available to plants in the form of nitrate (NO_3^-) or ammonia (NH_4^+) according to Chen et al. (2009). The first step in nitrogen assimilation is carried out by nitrate reductase, which uses nitrate as a substrate and reduces it to nitrite, which in turn is reduced to ammonia by nitrite reductase. Consequently, ammonia is used in the structure of amino acids (MIFLIN; LEA, 1976).

However, nitrogen is present in the plant in different forms. The greatest amount of this mineral is found in amino acids. Therefore, the higher the concentration of NO_3 - the greater the assimilation of nitrogen in the form of amino acids, favoring greater growth (CARDOSO et al., 2011).

Nitric oxide (NO) is another nitrogenous compound, derived from nitrate, which has been cited as an important signaling molecule in plants (GUPTA; ABIR, 2011; JIN et al., 2009).

Its formation comes from the reduction of nitrate to nitrite, which is reduced to nitric oxide in complexes III or IV or AOX of the electron transport chain, specifically when the concentration of the electron receptor, O_2, is low or absent. Another route of formation is through nitrate reductase, which reduces nitrate to nitric oxide (BETHKE; BADGER; JONES, 2004; GUPTA et al., 2011; LEA et al., 2004).

Liu et al. (2009) showed that the accumulation of NO in the first stage of soaking is beneficial for the rapid catabolism of ABA and consequent breaking of seed dormancy. This is due to the expression of the CYP707A2 gene, which participates in the catabolism of abscisic acid (ABA). However, Matakiadis et al. (2009) demonstrated that the CYP707A2 gene is also responsible for nitrate. In addition to these functions, nitrate, through nitric oxide, also stimulates the activity of superoxide dismutase (SOD), which is a key enzyme in the antioxidant system (PEREIRA et al., 2010).

2.4 Changes in the metabolism of macromolecules

The benefits provided by conditioning can occur for various reasons, but changes related to the content of macromolecules such as carbohydrates in general, proteins and amino acids are important events during osmotic conditioning (VARIER; VARI; DADLANI, 2010). Macromolecules can increase tolerance to stressful environments by maintaining the cell's osmotic potential, reducing Reactive Oxygen Species (ROS), maintaining the stability of membranes and proteins, as well as providing an alternative source of energy (BOHNERT; SHEN, 1999) and controlling the expression of various genes (GUPTA; KAUR, 2005).

According to Nawaz et al. (2011) there was an increase in the concentration of reducing, non-reducing and total soluble sugar in conditioned tomato seeds when compared to the control. Ismail et al. (2005) also observed an increase in the content of soluble sugar and amino acids. Braccini et al. (2000) observed in soybean seeds that storage causes a reduction in protein, but when these seeds underwent conditioning there was a decrease in this drop.

2.5 Antioxidant system

Another benefit of conditioning is the increased efficiency of the seed's antioxidant system. In terms of antioxidants, seeds have enzymatic and non-enzymatic defense systems (CHEN et al., 2009). The enzymatic system is made up of the enzymes catalase (CAT) and superoxide dismutase (SOD) (KIBINZA et al., 2011). The non-enzymatic system is made up of the ascorbic acid (AsA) cycle and reduced glutathione (GSH), with ascorbate peroxidase (APX), dehydroascorbate reductase (DHAR), monodehydroascorbate reductase (MDHA) and glutathione reductase (GR) being key enzymes in the cycle (MA; CHEN, 2004).

Each enzyme has a specific function, such as SOD, which acts to remove superoxide (O_2^-), converting it into hydrogen peroxide (H_2O_2), which can be converted into water and oxygen by CAT (KIBINZA et al., 2011). In the ASA-GR cycle, APX catalyzes the reduction of H_2O_2 into water with AsA serving as the electron donor (ZHANG et al., 2008). DHAR uses the electrons provided by GSH to reduce DHA into AsA, while DHA is previously produced from MDHA. Simultaneously, GSH is oxidized to glutathione disulfide (GSSG) by DHAR and GSSG is then reduced to GSH by the action of GR (GILL; TUTEJA, 2010).

The mitochondrial respiratory chain is an important source of Reactive Oxygen Species (ROS), which are present during periods of natural stress, such as germination, desiccation and ageing (KIBINZA et al., 2011; SUN et al., 2011). The imbalance between the production of ROS and their disappearance can lead to seed deterioration, a process known as oxidative stress (BAILLY, 2004). The main reactive species are the superoxide anion (O_2^-), hydrogen peroxide (H_2O_2), singlet oxygen ($^1O_2$) and the hydroxide radical ($\cdot OH$) (LI; LIU; ZANGH, 2010).

Improvements in the antioxidant system of the seeds were observed in conditioned seeds. Chen and Arora (2011), Kibinza et al. (2011) and Sun et al. (2011), together with other authors including Bailly et al. (2000) and Chiu, Chuang and Sung (2006), obtained a significant increase in the activity of the enzymes SOD, CAT, APX and the AsA-GSH cycle, as well as a decrease in MDA.

3 GENERAL CONSIDERATIONS

Osmotic conditioning is a technique that promotes various benefits to seeds, both metabolically and physiologically, enabling better germination performance. However, it is a technique that needs further study in order to define the best incubation media for certain species.

REFERENCES

ALBORESI, A. et al. Nitrate, a signal relieving seed dormancy in Arabidopsis. **Plant Cell Environment**, Oxford, v. 28, n. 4, p. 500-512, 2005.

ARIN, L. et al. Effects of different osmotic solutions on onion seed emergence. **African**

Journal of Agricultural Research, Ekpoma, v. 6, n. 4, p. 986-991, 2011.

BRAZILIAN SEED AND SEEDLING TRADE ASSOCIATION. **Abcsem participates in the IX course on vegetable seed production technology**. Available at: <http://www.abcsem.com.br/ noticia.php?cod=41>. Accessed on: Aug. 19, 2012.

BAiLLY, c. Active oxygen species and antioxidants in seed biology. **Seed Science Research**, Wallingford, v. 14, n. 2, p. 93-107, 2004.

BAiLLY, c. et al. Antioxidant systems in sunflower (*Helianthus annuus* L.) seeds as affected by priming. **Seed Science Research**, Wallingford, v.10, n. 1, p. 35-42, 2000.

BETHKE, P. c.; BADGER, M. R.; JoNEs, R. L. Apoplastic synthesis of nitric oxide by plant tissues. **Plant Cell**, Rockville, v. 16, n. 2, p. 332-341, 2004.

BOCIAN, S.; HOLUBOWICZ, R. Effect of different ways of priming tomato (*Lycopersicon esculentum* Mill.) seeds on their quality. **Journal of Natural Sciences**, olsztyn, v. 23, n. 4, p. 729-739, 2008.

BoHNERT, H. J.; sHEN, B. Transformation and compatible solutes. **Scientia Horticulturae**, Amsterdam, v. 78, p. 237-260, 1999.

BRACCINI, A. L. et al. Biochemical changes associated with soybean seeds osmoconditioning during storage. **Pesquisa Agropecuària Brasileira**, Brasilia, v. 35, n. 2, p. 433-447, 2000.

BRADFORD, K. J. Manipulation of seed water relations via osmotic priming to improve germination under stress conditions. **HortScience,** Alexandria, v. 21, n. 5, p.1105-1112, 1986.

CARDOSO, S. M. et al. Nitrogen sources and installments in cover crops of corn under no-till. **Revista Brasileira de Ciências Agràrias**, Recife, v. 6, n. 1, p. 23-28, 2011.

CASTRO, R. D. et al. Cell division and subsequent radicle protrusion in tomato seeds are inhibited by osmotic stress but DNA synthesis and formation of microtubular cytoskeleton are not. **Plant physiology**, Minneapolis, v. 122, n. 2, p. 327-336, 2000.

CHEN, K.; ARORA, R. Dynamics of the antioxidant system during seed osmopriming, post-priming germination, and seedling establishment in Spinach (*Spinacia oleracea*). **Plant science**, Limerick, v. 180, n. 2, p. 212-220, 2011.

CHEN, L. et al. Effects of nitrogen forms on the growth , ascorbate-glutathione cycle and lipid peroxidation in developing seeds of vegetable soybean. **Journal of Agricultural Research**, Washington, v. 4, n. 11, p. 1178-1188, 2009.

CHIU, K. Y.; CHUANG, S. J.; SUNG, J. M. Both anti-oxidation and lipidcarbohydrate conversion enhancements are involved in priming-improved emergence of *Echinacea purpurea* seeds that differ in size. **Scientia Horticulturae**, Amsterdam, v. 108, n. 2, p. 220-226, 2006.

CORTEZ-BAHEZA, C. E. et al. A new LEA gene is induced during osmopriming of *Capsicum annuum* L. seeds. **International Journal of Botany**, New York, v. 4, n. 1, p. 99-106, 2007.

EL-ARABY, M. M.; HEGAZI, A. Z. Responses of tomato seeds to hydro- and osmo-priming, and possible relations of some antioxidant enzymes and endogenous polyamine fractions. **Egyptian Journal of Biology,** Ismailia, v. 6, p. 81-93, 2004.

FAROOQ, M. et al. Changes in nutrient-homeostasis and reserves metabolism during rice seed priming: consequences for seedling emergence and growth. **Agricultural Sciences in China**, Beijing, v. 9, n. 2, 191-198, 2010.

FOOD AND AGRICULTURAL ORGANIZATION. **Country information**: Brazil: agriculture sector. 2007. Available at: <http://www.fao.org>. Accessed on: Aug. 21, 2012.

GILL, S. S.; TUTEJA, N. Reactive oxygen species and antioxidant machinery in abiotic stress tolerance in crop plants. **Plant physiology and Biochemistry**, Paris, v. 48, n. 12, p. 909-930, 2010.

GOVINDEN-SOULANGE, J.; LEVANTARD, M. Comparative studies of seed priming and pelleting on percentage and meantime to germination of seeds of tomato (Lycopersicon *esculentum* Mill.). **African Journal of Agricultural Research**, Ekpoma, v. 3, n. 10, p. 725-731, 2008.

GUPTA, K. J.; ABIR, U. I. The anoxic plant mitochondrion as a nitrite: no reductase. **Mitochondrion**, Birmingham, v. 11, n. 4, p. 537-43, 2011.

GUPTA, K. J. et al. The emerging roles of nitric oxide (NO) in plant mitochondria. **Plant Science**, Limerick, v. 181, n. 5, p. 520-526, 2011.

HEYDECKER, W.; GIBBINS, B. M. The priming of seeds. **Acta Horticulturae**, The Hague, v. 83, p. 213-223, 1978.

BRAZILIAN INSTITUTE OF GEOGRAPHY AND STATISTICS. 2011.

Available at: <http://www.ibge.gov.br/home>. Accessed on: 29 Jan. 2013.

ISMAIL, A. I. et al. Optimization of priming benefits in tomato (Lycopersiconesculentum M.) and changes in some osmolytes during the hydration phase. **Asian Journal of Plant Sciences**, New York, v. 4, n. 6, p. 691701, 2005.

JIN, C. W. et al. Differential regulatory role of nitric oxide in mediating nitrate reductase activity in roots of tomato (*Solanumly cocarpum*). **Annals of Botany**, London, v. 104, n. 1, p. 9-17, 2009.

KIBINZA, S. et al. Catalase is a key enzyme in seed recovery from ageing during priming. **Plant science**, Limerick, v. 181, n. 3, p. 309-315, 2011.

KISSMANN, C. et al. Bioregulator and osmotic preconditioning on seed germination and initial growth of carob seedlings. **Revista Brasileira de Plantas Medicinais**, Botucatu, v. 13, n. 1, p. 58-67, 2011.

LANTERI, S. et al. The effects of priming on nuclear replication activity and germination of pepper (*Capsicumannuum*) and tomato

(*Lycopersiconesculentum*) seeds. **Seed Science Research,** Wallingford, v. 4, n. 2, p. 81-87, 2008.

LEA, U. S. et al. Mutation of the regulatory phosphorylation site of tobacco nitrate reductase results in high nitrite excretion and NO emission from leaf and root tissue. **Planta**, Berlin, v. 219, n. 1, p. 59-65, 2004.

LIU, Y. et al. Nitric oxide-induced rapid decrease of abscisic acid concentration is required in breaking seed dormancy in Arabidopsis. **The New phytologist**, Cambridge, v.183, n. 4, p.1030-42, 2009.

LI, Y.; LIU, Y.; ZHANG J. Advances in the research on the AsA-GSH cycle in horticultural crops. **Frontiers of Agriculture in China**, Baoding, v. 4, n. 1, p. 84-90, 2010.

MATAKIADIS, T. et al. The Arabidopsis abscisic acid catabolic gene CYP707A2 plays a key role in nitrate control of seed dormancy. **Plant Physiology,** Minneapolis, v. 149, n. 2, p. 949-960, 2009.

MIFLIN, B. J.; LEA, P. J. The pathway of nitrogen assimilation in plants. **Phytochemistry**, New York, v. 15, n. 6, p. 873-885, 1976.

NASCIMENTO, W. M. **Osmotic conditioning of vegetable seeds** *5*. Brasilia: Embrapa Hortaliças, 2004. 12 p. (Technical Circular, 33). NAWAZ, A. et al. Effect of halopriming on germination and seedling vigor of tomato. **African Journal of Agricultural Research**, Ekapoma, v. 6, n. 15, p. 3551-3559, 2011.

NAWAZ, A. et al. Induction of salt tolerance in tomato (*Lycoper5icon e5culentum* Mill .) seeds through sand priming. **Australian Journal of Crop Science**, Lismore, v. 6, n. 7, p. 1199-1203, 2012.

PACE, R. et al. Germination of untreated and primed seeds in rapeseed (*Bra55ica napu5* var. Oleifera DEL.) under salinity and low matric potential. **Experimental Agriculture**, Cambridge, v. 48, n. 2, p. 238-251, 2012.

PEREIRA, B. L. C. et al. Influence of nitric oxide on the germination *of Plathymenia reticulata* Benth seeds with low vigor. **Scientia Forestalis**, Piracicaba, v. 38, n. 88, p. 629-636, 2010.

POSSE, S. C. P. et al. Effects of osmotic conditioning and hydration on the twinning of pepper seeds (*Capsicum annuum* L.) submitted to low temperatures. **Revista Brasileira de Sementes**, Brasilia, v. 23, n. 1, p. 123-127, 2002.

SUN, H. et al. Ascorbate-glutathione cycle of mitochondria in osmoprimed soybean cotyledons in response to imbibitional chilling injury. **Journal of Plant Physiology,** Minneapolis, v. 168, n. 3, p. 226-32, 2011.

VARIER, A.; VARI, A. K.; DADLANI, M. The subcellular basis of seed priming. **Current Science**, Bangalore, v. 99, n. 4, p. 450-456, 2010.

ZHANG, Y. et al. Chilling acclimation induced changes in the distribution of H2O2 and antioxidant system of strawberry leaves. **Agricultural Journal,** Bridgetown, v. 3, n. 4, p. 286-291, 2008.

ARTICLE 1 CHANGES IN THE ENZYMATIC PATTERN OF TOMATO SEEDS CAUSED BY OSMOTIC CONDITIONING

SUMMARY

Osmotic conditioning is a technique used to improve the germination performance of tomato seeds. The aim was to study the influence of the conditioning agents PEG, KNO3 or both on the activity of some enzymes involved in germination metabolism and on the physiological performance of conditioned tomato seeds. Tomato seeds (*Lycopersicon esculentum* Mill.) cv Santa Clara were conditioned in solutions of polyethylene glycol 6000 (PEG) and potassium nitrate (KNO_3) for 2, 4 and 6 days. The osmotic agent solutions (v/v) were: 100% PEG for 2, 4 and 6 days; 100% KNO_3 for 2, 4 and 6 days; and combinations of 50% PEG + 50% KNO_3 for 2, 4 and 6 days. After conditioning, biochemical and physiological analyses were carried out. In the conditionings with 100% (KNO_3) for 4 and 6 days, high germination speed index (GVI), seedling formation speed index (SPI) and emergence speed index (ESI) were observed due to the higher activity of nitrate reductase, which possibly acted to increase nitrogen assimilation and also to increase the activity of enzymes in the antioxidant system. Therefore, conditioning in 100% KNO3 for 4 and 6 days increases the activity of nitrate reductase and superoxide dismutase, contributing to high speed indices in tomato seeds.

Keywords: Nitrate reductase. Osmotic conditioning. Superoxide dismutase seed.

1 INTRODUCTION

The estimated tomato yield in 2012 for Brazil was approximately 4 million tons in a planted area of around 66,000 hectares (INSTITUTO BRASILEIRO DE GEOGRAFIA E ESTATiSTICA - IBGE, 2012). It is a vegetable of great importance to the economy, whose values are due to the high productivity obtained, among other factors, due to the quality of the seeds.

Currently, techniques that promote improvements in seeds aim to increase vigor, which has been satisfactorily obtained through osmotic conditioning. This technique promotes faster germination and uniformity in batches (NASCIMENTO, 2004), and is being used frequently on tomato seeds (BOCIAN; HOLUBOWICZ, 2008; NAWAZ et al., 2012; ROSSETTO; LIMA; NAKAGAWA, 2002).

Osmotic conditioning involves the seeds absorbing water in a controlled manner, using compounds to control the potential of the solution (HEYDECKER; GIBBINS, 1978). As a result, germination processes only begin, without germination taking place.

Among the compounds used are the inert ones (polyethylene glycol and mannitol), which due to their high molecular weight are not absorbed by the seeds, but the high viscosity of the solution can make aeration difficult. In addition to saline solutions (potassium nitrate, sodium chloride, etc.), which can be absorbed by the seeds and have low viscosity, they are also easy to aerate (NASCIMENTO, 2004).

The absorption of certain ions can influence germination by activating enzymes in metabolic pathways that would not be active naturally (NAWAZ et al., 2012; PACE et al., 2012). Among these possible ions, nitrate deserves further study because it is used on a large scale in tomato seeds and most often comes from the dissociation of potassium nitrate (BOCIAN; HOLUBOWICZ, 2008; MAVI; ERMIS; DEMIR, 2006; NASCIMENTO, 2004). Furthermore, little is known about the mechanisms that act on seeds when they absorb these ions during conditioning and the effects on metabolism.

Therefore, studying the activity of the nitrate reductase enzyme may be important in elucidating metabolic issues in tomato seeds conditioned in potassium nitrate solution. This enzyme is essential for nitrogen assimilation, as it converts nitrate (NO_3^-) into nitrite (NO_2^-) (MIFLIN; LEA, 1976). Nitrate reductase has been extensively studied in plant organs, but there has been little work on seeds. In addition, according to Pereira et al. (2010) another factor related to nitrate and the metabolism of conditioned seeds is the stimulation of the activity of the enzyme superoxide dismutase (SOD), which together with catalase (CAT) and ascorbate peroxidase (APX), play a role in the removal of reactive oxygen species (ROS) (CHEN; ARORA 2011).

In the removal of EROS, superoxide anion (O_2^-) is removed by SOD, which catalyzes its conversion to hydrogen peroxide (H O_{22}), which in turn can be reduced to water and oxygen by CAT. APX catalyzes hydrogen peroxide (H_2O_2) using the AsA-GSH cycle (CHEN; ARORA 2011; TULLIO; ARRIGONI, 2003).

The aim was to study the influence of the conditioning agents PEG, KNO3 or both on the activity of some enzymes involved in germination metabolism and on the physiological performance of tomato seeds.

2 MATERIAL AND METHODS

Tomato seeds (*Lycopersicon esculentum* Mill.) cv Santa Clara were conditioned in solutions of polyethylene glycol 6000 (PEG) and potassium nitrate (KNO_3) for 2, 4 and 6 days. The osmotic agent solutions (v/v) were: 100% PEG for 2, 4 and 6 days; 100% KNO_3 for 2, 4 and 6 days; and combinations of 50% PEG + 50% KNO_3 for 2, 4 and 6 days. All the solutions had a Ψs of - -1.1 MPa, equivalent to a concentration of 0.05M PEG and 0.22M KNO_3 . The seeds were incubated under constant light and aeration, at a temperature of 15 °C, with 15 mL of solution per gram of seed. Osmotic potential calculations were obtained using the Michel and Kaufmann equation (1973) for PEG6000 and the Van'Thoff equation (HILLEL, 1971) for KNO_3 . The control was not conditioned. After conditioning, the seeds were washed under running water to remove the osmoconditioning solution from the surface and dried in a room with a temperature of 25 °C and relative humidity of 50% for a period of twenty-four hours, when the seeds reached a humidity level of between 5-7%.

Electrical conductivity was measured using the *bulk* system, according to the methodology proposed by the vigor committee of the *Association of Official Seed Analysts* - AOSA (1983), described by Marcos Filho, Cicero and Silva (1987). Four replicates of 50 seeds were used, conditioned in 100%PEG for 6 days, 50%PEG + 50%KNO3 for 6 days and 100%KNO3 for 6 days, washed in running water and dried to initial humidity, then weighed and placed in plastic cups containing 75 mL of deionized water. The cups were kept in a germinator at 25 °C and readings were taken after 24 hours of soaking, with the results expressed as $\mu S\ cm^{-1} \cdot g^{-1}$.

Sulphuric digestion was used to quantify total nitrogen (*ASSOCIATION OF OFFICIAL ANALYTICAL CHEMISTRY* - AOAC, 1990). To do this, 10 mg of seed were weighed and transferred to a digestion tube, adding 1.5 g of potassium sulphate (K_2SO_4) and 0.3 g of copper sulphate ($CuSO_4$) plus 3.0 mL of concentrated sulphuric acid (H_2SO_4). The total nitrogen content was determined according to the *Kjeldahl* method.

The extracts for the analysis of total proteins and amino acids were obtained by macerating 200 mg of seeds in 5 mL of potassium phosphate buffer (100 mM; pH 7.0) and submitting them to a water bath at 40^0 C for 30 minutes, centrifuged at 10,000 rpm for 10 minutes. After collecting the supernatant, the precipitate was resuspended in 5 mL of potassium phosphate buffer (100 mM; pH 7.0) and the supernatant was collected again and added to the first.

Protein and amino acid quantifications were determined colorimetrically using the Bradford method (1976) and the Ninhydrin method (YEMM; COCCKING, 1955), respectively. For both analyses, aliquots of 30 μL of the extract were used.

To determine the activities of nitrate reductase and enzymes of the antioxidant system, the conditioned seeds and the control were soaked in distilled water for 24 hours. Nitrate reductase activity was measured according to Delù Filho, Oliveira and Alves (1997), with some modifications. 600mg of whole seeds *in vivo were* used, according to the pre-test, which were vacuum infiltrated for two minutes and then subjected to a water bath for 75 minutes. Finally, a 2 mL aliquot was used. The concentration of nitrite produced was determined using a spectrophotometer at 540 nm.

To extract the antioxidant enzymes, 200 mg of seeds were macerated in liquid N_2 and homogenized in a buffer with the following composition: 100 mM potassium phosphate, pH 7.8, 0.1 mM EDTA and 1 mM ascorbic acid. The homogenates were centrifuged at 13,000 g for 20 minutes at 4°C and the supernatants were collected for enzymatic analysis of superoxide dismutase (SOD), catalase (CAT) and ascorbate peroxidase (APX) (BIEMELT; KEETMAN; ALBRECHT, 1998).

SOD activity was assessed by the enzyme's ability to inhibit the photoreduction of nitrotetrazolium blue (NBT) (GIANNOPOLITISE; RIES, 1977) in an incubation medium composed of 50 mM potassium phosphate buffer, pH 7.8, 14 mM methionine, 0.1 μM EDTA, 75 μM NBT and 2 μM riboflavin. The tubes containing the reaction medium and the sample were illuminated for 7 minutes with a 20W fluorescent lamp. For the control, the same reaction medium without the sample was illuminated and the blank was kept in the dark. The readings were taken at 560 nm and the enzyme activity was calculated using the following equation: % inhibition = (A_{560} sample with enzyme extract - A_{560} control without enzyme) / (A_{560} control without enzyme). One unit of SOD corresponds to the amount of enzyme capable of inhibiting NBT photoreduction by 50% under the test conditions.

CAT was assessed by the decrease in absorbance at 240 nm for 3 minutes in an incubation lid containing 200 mM potassium phosphate, pH 7.0 and 12.5 mM H O_{22} , incubated at 28° C, in which the consumption of hydrogen peroxide was monitored (HAVIR; MCHALE, 1987).

APX activity was measured according to Nakano and Asada (1981), monitoring the rate of ascorbate oxidation at 290 nm. The incubation buffer was composed of 50 mM potassium phosphate, pH 7.0, 0.5 mM ascorbic acid and 0.1 mM H O .22

The speed of germination (IVG), normal seedling formation (IVP) and emergence (IVE) indices were based on three replicates of 50 seeds. To obtain the first and second indices, the seeds were placed on a paper substrate moistened with distilled water equivalent to 2.5 times the weight of the dry paper. To obtain the IVE, the seeds were germinated in *gerboxes* containing washed sand substrate. Both germination tests were maintained at an alternating temperature of 20-30 °C (optimum temperature) and a 12-hour photoperiod, according to Brasil (2009). Germination, normal seedling formation and emergence were counted daily to calculate the following indices: germination velocity index (GVI), normal seedling formation velocity index (NPVI) and emergence velocity index (EVI) according to Maguire (1962), with some modifications to the methodology, replacing the number of germinated seeds with the number of normal seedlings and emergence to calculate their respective indices.

The experimental design of the tests was DIC (Completely Randomized Design), following the factorial scheme (3x3) +1, where: 3 osmotic solutions (100% PEG, 50%PEG+50%KNO3 and 100% KNO3); 3 incubation times (2, 4, 6 days) plus the control (unconditioned seeds). Only for the conductivity test was the 3x1 + 1 factorial scheme used: 3 solutions (100% PEG; 50%PEG+50%KNO3; 100%KNO3); 1 period (6 days); plus the unconditioned control. Statistical analysis was carried out using the *Sisvar* statistical program (FERREIRA, 1999) and the means were compared using the *Scott-Knott* test at a 5% probability of error ($P < 0.05$). The graphs were constructed using the *SigmaPlot® software.*

3 Results and discussion

The electrical conductivity observed in tomato seeds conditioned in 100% KNO_3 for 6 days was higher than in the other solutions (Table 1). This result is similar to that found by José et al. (1999), who observed lower electrical conductivity in PEG treatments compared to KNO3. This higher conductivity may be associated with the ease with which tomato seeds absorb the ions K^+ and NO_3 - (SANTOS et al., 2008) and there is no greater destabilization of the plasma membrane, since seeds conditioned with 100% KNO_3 for 6 days showed high velocity indices.

Although there was no statistically significant difference in the concentration of total nitrogen, conditioning with 100% KNO3 for 4 and 6 days showed a tendency towards a higher concentration of total nitrogen (Figure 1), contributing to the hypothesis of ion absorption.

Table1 Electrical conductivity in conditioned and unconditioned tomato seeds - UFLA - 2013.

TREATMENTS	COND. ELECTRICAL
100%PEG/6 days	23,07 b
50%PEG+50%KNO3/6 days	29,80 b
Witness	31,73 b
100%KNO3/6 days	43,26 a
cv (%)	22,03

Means followed by the same letters in the columns do not differ by the *Scott-Knott* test at 5% probability. Data in (µS cm-[1] g-).[1]

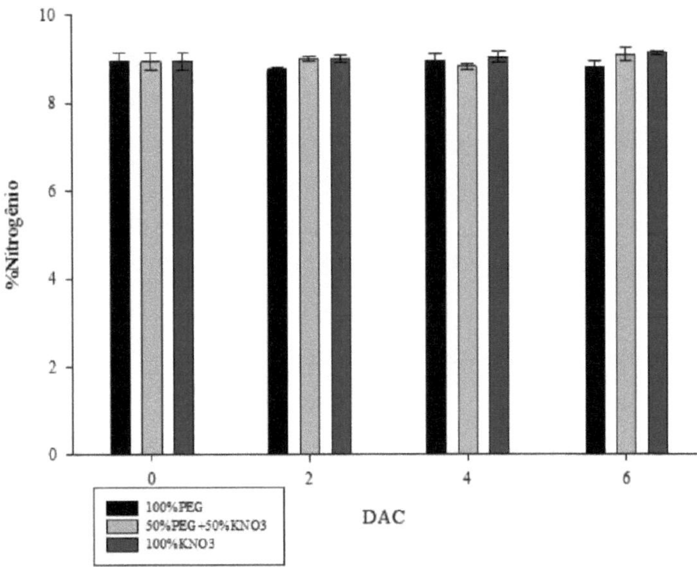

Figura 1 Percentage of total nitrogen in conditioned and unconditioned tomato seeds. The bars represent the standard deviation of the means.

The highest nitrate reductase (NR) activities were observed in seeds conditioned in 100% KNO solutions3 . In which 100% KNO3 conditioning for 6 days increased NR activity by 70% compared to the control, followed by 100% KNO_3 conditioning for 4 and 2 days, which increased activity by 60% compared to the control. No statistical difference was observed in the other solutions (Figure 2). However, there was a difference in NR activity, even though there was no statistical difference in total nitrogen concentrations (Figure 1)

The increase in NR activity in seeds conditioned with KNO3 was stimulated by the increase in the concentration of NO ions$_{3}^{-}$, which come from the dissociation of KNO3. These are a substrate for the enzyme (MIFLIN; LEA, 1976). Thus, RN proved to be more active in tomato seeds when substrate availability increased and, consequently, conferred positive increases in the physiological variables, suggesting a positive relationship between reductase activity and the physiological variables studied. Therefore, the entry of certain ions into the seeds would be important for the activation of metabolic pathways that are inactive under natural conditions, which could lead to better performance than genetically expected.

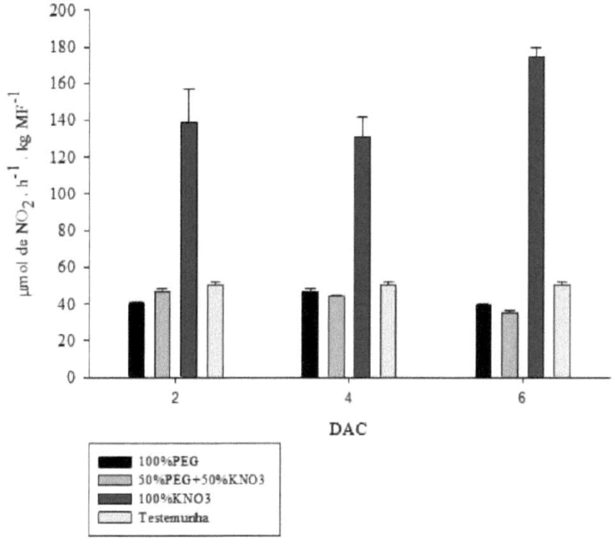

Figura 2 Nitrate reductase activity in tomato seeds subjected to different conditionings and non-conditionings. The bars represent the standard deviation of the means.

It can be seen that the seeds from the 100%PEG 2 and 6 day treatments, as well as the control, had the lowest total protein content, while the 100%PEG 4 day treatment did not differ from the other treatments, maintaining the highest total protein concentrations (Figure 3). The increase in total protein content in seeds during the beginning of the germination process is associated with the induction of the biosynthesis of hydrolytic enzymes used in germination (PONTES et al., 2002).

Total amino acids did not differ statistically, remaining between 0.008 and 0.012g.g.MS^{-1} .

This may not be a very accurate technique for identifying small variations in amino acid concentrations, because as well as differences in total protein, there are differences in RN activity.

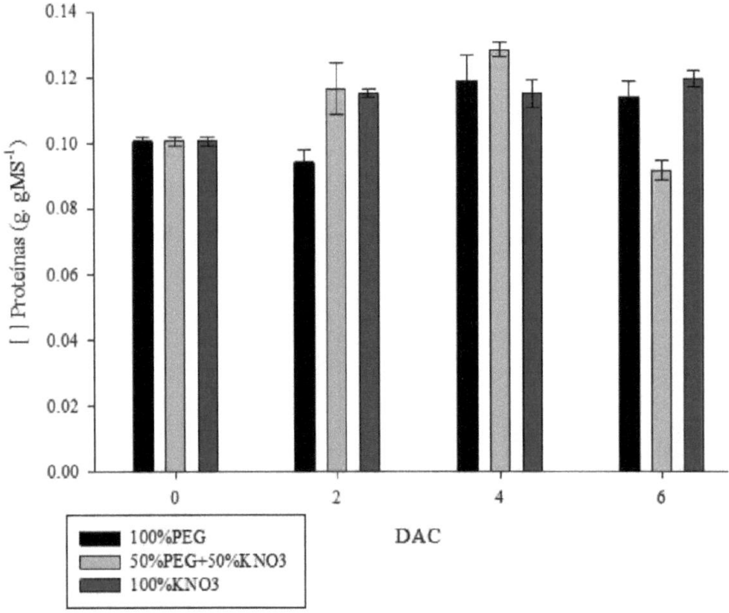

Figura 3 Amount of total protein in tomato seeds subjected to different conditionings and non-conditionings. The bars represent the standard deviation of the means.

Conditioning in 100% KNO3 for 4 and 6 days increased SOD activity by around 40% compared to the control. However, the other conditions did not show any differences compared to the control.

These results differ from those found in the literature (CHEN; ARORA, 2011; WEITING; TRELEASE; EISING, 1990) which observed lower SOD activity in seeds in the transition phase between quiescence and germination for conditioned seeds. The observed increase may have been caused by stimulation from nitric oxide (NO) (PEREIRA et al., 2010) formed by NO_3^- absorbed during conditioning. According to Bethke, Badger and Jones (2004) and Gupta and Abir (2011), NO is formed by nitrite (NO_2^-) or by the activity of nitrate reductase itself (LEA et al., 2004). However, it is believed that NO_3^- can be formed from NO.

Thus, the seeds conditioned in 100% KNO_3, 4 days and 6 days with the highest nitrate reductase activity (Figure 2), and consequently the highest nitrite formation, were those in which the highest SOD activity was observed. With the exception of those conditioned in 100% KNO3 for 2 days, which showed high NR activity but low SOD activity. This is possibly due to the shorter time it takes for NO to form, as Pereira et al. (2010) in their study observed that it took approximately 48 hours for NO to form from NO_3^- . However, studies aimed at quantifying whether NO is formed and whether it comes from NO^- 3 resulting from reductase activity in the conditioned seeds need to be carried out.

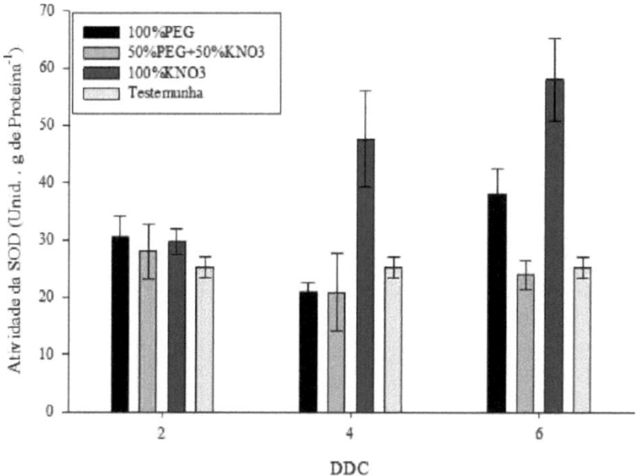

Figura 4 Superoxide dismutase (SOD) activity in tomato seeds subjected to different conditioning and non-conditioning. The bars represent the standard deviation of the means.

Catalase activity was not observed at 2 days of conditioning for any of the solutions tested. In addition, conditioning in 50%PEG+50%KNO3 for 4 days and 6 days showed lower activity compared to the control. Only seeds conditioned in 100%KNO3 for 6 days showed a 15% increase in catalase activity compared to the control. The other seeds did not differ from the control (Figure 5). The 100% KNO3 6-day conditioning gave this result in response to the high activity of SOD, whose product is H_2O_2, which is the catalase substrate (CHEN; ARORA, 2011). Thus, the increase in H2O2 concentration stimulated catalase.

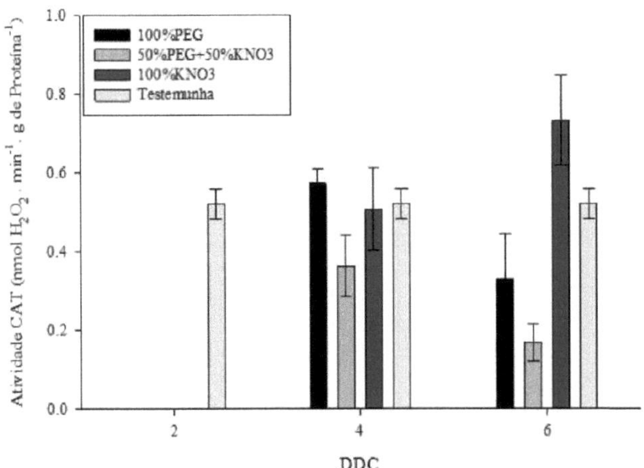

Figura 5 Catalase activity (CAT) in tomato seeds subjected to different conditioning and non-conditioning. The bars represent the standard deviation of the means.

APX activity in seeds conditioned in 100% KNO3 for 4 and 6 days did not differ from the control. The activity of the other solutions was at least 50% higher when compared to the control. Conditioning in 100%PEG for 4 and 6 days increased activity by around 65% when compared to the control (Figure 6). Similar results were found by Chen and Arora (2011) who observed higher APX activity in seeds conditioned with PEG. The low activity shown by the control batch is related to the transition phase from quiescence to germination, in which other enzymes of the antioxidant system play a more significant role (TULLIO; ARRIGONI, 2003).

The low APX activity observed in the 100%KNO3 4 and 6 day conditioning, in relation to the other solutions, may be due to the use of another route to dissociate H2O2, because as observed, the high activity of CAT in these seeds. In general, seeds conditioned in 100% KNO3 for 4 days and 6 days used CAT to dissociate H2O2. Due to the high activity of SOD in these seeds, it is possible that there is a high production of H O22 , with catalase being responsible for removal in this case. Seeds conditioned with 100%PEG 4 days and 6 days use APX to dissociate H O22 , as it removes small concentrations due to its greater affinity for it (MITTLER, 2002; SHIGEOKA et al., 2002). This relationship can also be seen in relation to conditioning with 50%PEG+50%KNO3 for 4 days and 6 days, which show low CAT activity and high APX activity. In addition, according to Sun et al. (2011) seeds conditioned with PEG after 72 hours of conditioning have a decrease in H2O2 content of around 60% compared to the control.

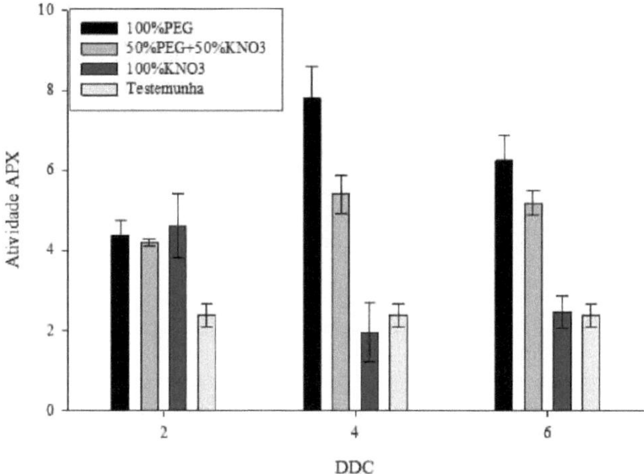

Figura 6 Ascorbate peroxidase (APX) activity in tomato seeds subjected to different conditioning and non-conditioning. The bars represent the standard deviation of the means.

Conditioning in 100%KNO3 6 days and 50%PEG+50%KNO3 4 days increased the GVI by around 43% compared to the control and 100%KNO3 4 days increased it by around 30% (Table 2). *The* rate of normal seedling formation (IVP) was higher in seeds conditioned in 100%KNO3 4 days and 6 days, 50%PEG+50%KNO3 4 days and 100%PEG 4 days (Table 2). Again, conditioning in 100%KNO3 2 days, 4 days and 6 days, together with 100%PEG 4 days, increased the IVE by around 20% compared to the control (Table 2). In addition, seeds conditioned for 4 days in any of the solutions tested had higher rates than the control.

Table 2 Velocity indices of conditioned and unconditioned tomato seeds - UFLA - 2013.

TREATMENTS	IVG	IVP	IVE
Witness	12.06 d	5.43 b	3.24 b
100%PEG/2 days	12.78 d	5.35 b	3.51 b
100%PEG/4 days	17.00 c	6.53 a	4.56 a
100%PEG/6 days	15.22 c	5.90 b	3.35 b
50%PEG+50%KNO3/2 days	13.55 d	5.61 b	3.23 b
50%PEG+50%KNO3/4 days	21.70 a	7.26 a	3.65 b

50%PEG+50%KNO3/6 days	16.40 c	6.17 b	3.47 b
100%KNO3/2 days	14.95 c	5.87 b	4.09 a
100%KNO3/4 days	17.86 b	6.36 a	3.99 a
100%KNO3/6 days	21.34 a	6.75 a	4.36 a
CV (%)	8.59	14.38	12.71

GVI = germination rate; NPVI = normal seedling formation rate; ESI = emergence rate. Means followed by the same letters in the columns do not differ by the Scott-Knott test at 5% probability.

Conditioning in 100% KNO3 for 4 days and 6 days generally increased the speed indices studied. This result is similar to those observed by Bocian and Holubowicz (2008) who worked with tomatoes and Ahmadvand et al. (2012) and Reis et al. (2012) who worked with soybeans and eggplants, respectively, who observed a better physiological performance of seeds conditioned with KNO_3 compared to PEG. However, as observed in this study, the better physiological performance of tomato seeds conditioned in KNO_3 is related to an increase in nitrate reductase activity. This is due to the greater availability of substrate provided by the KNO3 solution. In addition, nitrate reductase may be indirectly related to an increasc in SOD and CAT enzyme activity through nitric oxide, but further studies are needed to validate this hypothesis. However, the role of nitrate reductase has not yet been investigated in this situation.

The high speed indexes provided by the 50%PEG+50%KNO3 4-day and 100%PEG 4-day treatments may be related to the concentration of total proteins, which can be used as a reserve of small peptides and amino acids, which can later supply organic nitrogen during initial development (LIMA et al., 2008). Another advantage of conditioning with 100%PEG for 4 days is the increase in CAT activity, which is a key enzyme in protecting seeds against the accumulation of H O_{22} , thus reducing oxidative stress and favoring germination (KIBINZA et al., 2011).

4 CONCLUSION

Conditioning in 100% KNO3 for 4 and 6 days gave the highest nitrate reductase and superoxide dismutase activities, as well as the highest germination, normal seedling formation and emergence rates.

REFERENCES

AHMADVAND, G. et al. Effect of seed priming with potassium nitrate on germination and emergence traits of two soybean cultivars under salinity stress conditions. **American-Eurasian Journal of Agricultural & Environmental Science**, Dubai, v. 12, n. 6, p. 769-774, 2012.

ASSOCIATION OF OFFICIAL ANALYTICAL CHEMISTRY. **Official methods of analysis of the Association of Official Analytical Chemists.** 14th ed. Arlington, 1984. 1141 p.

ASSOCIATION OF OFFICIAL SEED ANALYSTS. **Seed vigour testing handbook.** Lincoln, 1983. 88 p.

BETHKE, P. C.; BADGER, M. R.; JONES, R. L. Apoplastic synthesis of nitric oxide by plant tissues. **Plant Cell**, Rockville, v. 16, n. 2, p. 332-341, 2004.

BIEMELT, S.; KEETMAN, U.; ALBRECHT, G. Re-aeratio following hypoxia or anoxia leads to activation of the antioxidative defense in roots of wheat seedlings. **Plant Physiology,** Minneapolis, v. 116, n. 2, p. 651-658, 1998.

BOCIAN, S.; HOEUBOWICZ, R. Effect of different ways of priming tomato (*Lycopersicon esculentum* Mill.) seeds on their quality. **Journal of Natural Sciences**, Olsztyn, v. 23, n. 4, p. 729-739, 2008.

BRADFORD, M. M. A rapid and sensitive method for the quantitation of microgram quantities of protein utilizing the principle of protein-dye binding.

Analytical Biochemistry, New York, v. 72, p. 248-254, 1976.

BRAZIL. Ministry of Agriculture, Livestock and Supply. **Rules for seed analysis.** Brasilia, 2009. p. 395.

CHEN, K.; ARORA, R. Dynamics of the antioxidant system during seed osmopriming, post-priming germination, and seedling establishment in Spinach (*Spinacia oleracea*). **Plant science**, Limerick, v. 180, n. 2, p. 212-220, 2011.

DELÙ FILHO, N.; OLIVEIRA, L. E. M.; ALVES, J. D. Nitrate reductase activity in young rubber tree plants (*Hevea brasiliensis Muell*) optimization of test conditions and circadian rhythm. **Revista Arvore**, Viçosa, MG, v. 21, n. 3, p. 329-336, 1997.

FERREIRA, D. F. **Sisvar**: statistical analysis system. Lavras: UFLA, 1999.

GIANNOPOLITIS, C. N.; RIES, S. K. Superoxide dismutase: I. Occurrence in higher plants. **Plant Physiology**, Minneapolis, v. 59, n. 2, p. 309-314, 1977.

GUPTA, K. J.; ABIR, U. I. The anoxic plant mitochondrion as a nitrite:NO reductase. **Mitochondrion**, Birmingham, v. 11, n. 4, p. 537-43, 2011.

HAVIR, E. A.; MCHALE, N. A. Biochemical and developmental characterization of multiple forms of catalase in tobacco leaves. **Plant Physiology**, Minneapolis, v. 84, n. 2, p. 450-455, 1987.

HEYDECKER, W.; GIBBINS, B. M. The priming of seeds. **Acta Horticulturae**, The Hague, v. 83, p. 213-223, 1978.

HILLEL, D. The state of water in the soil. In:. **Soil and water:** physical principles and processes. New York: Academic, 1971. p. 49-77.

BRAZILIAN Institute of Geography and STATISTICS. Available at: <http://www.ibge.gov.br/home/estatistica/indicadores/agropecuaria/lspa/ lspa_201202.pdf>. Accessed on: September 23, 2012.

JOSÉ, S. C. B. R. et al. Physiological and biochemical changes in pepper seeds submitted to osmoconditioning using different osmotic agents and soaking media. **Revista Brasileira de Sementes**, Brasilia, v. 21, n. 2, p. 217-223, 1999.

KIBINZA, S. et al. Catalase is a key enzyme in seed recovery from ageing during priming. **Plant science**, Limerick, v. 181, n. 3, p. 309-315, 2011.

LEA, U. S. et al. Mutation of the regulatory phosphorylation site of tobacco nitrate reductase results in high nitrite excretion and NO emission from leaf and root tissue. **Planta,** Berlin, v. 219, n. 1, p. 59-65, 2004.

LIMA, R. B. S. et al. Primary metabolite mobilization during germination in rosewood (*Aniba rosaeodora* Ducke) seeds. **Revista Arvore**, Viçosa, MG, v. 32, n. 1, p. 19-25, 2008.

MAGUIRE, J. D. Speed of germination-aid in relation evaluation for seedling emergence vigor. **Crop Science**, Madison, v. 2, n. 2, p. 176-177, 1962.

MARCOS FILHO, J.; CiCERO, S. M.; SILVA, W. R. **Avaliaçâo da qualidade fisiológica das sementes**. Piracicaba: FEALQ, 1987. 230 p.

MAVI, K.; ERMIS, S.; DEMIR, I. The effect of priming on tomato rootstock seeds in relation to seedling grofth. **Asian Journal of Plant Sciences**, New York, v. 5, n. 6, p. 940-947, 2006.

MIFLIN, B. J.; LEA, P. J. The pathway of nitrogen assimilation in plants.

Phytochemistry, New York, v. 15, n. 6, p. 873-885, 1976.

MITTLER, R. Oxidative stress, antioxidants and stress tolerance. **Trends in Plant Science**, Oxford, v. 7, n. 9, p. 405-410, 2002.

NAKANO, Y.; ASADA, K. Hydrogen peroxide is scavenged by ascorbate specific peroxidase in spinach chloroplasts. **Plant Cell Physiology**, Oxford, v. 22, n. 5, p. 867-880, 1981.

NASCIMENTO, W. M. **Osmotic conditioning of vegetable seeds**. Brasilia: Embrapa, 2004. p. 12

NAWAZ, A. et al. Induction of salt tolerance in tomato (*Lycopersicon esculentum* Mill .) seeds through sand priming. **Australian Journal of Crop Science**, Lismore, v. 6, n. 7, p. 1199-1203, 2012.

PACE, R. et al. Germination of untreated and primed seeds in rapeseed (*Brassica napus* var. Oleifera DEL.) under salinity and low matric potential. **Experimental Agriculture**, Cambridge, v. 48, n. 2, p. 238-251, 2012.

PEREIRA, B. L. C. et al. Influence of nitric oxide on the germination *of Plathymenia reticulata* Benth seeds with low vigor. **Scientia Forestalis**, Piracicaba, v. 38, n. 88, p. 629-636, 2010.

PONTES, C. A. et al. Reserve mobilization in seeds of *Apuleia leiocarpa* (Vogel). J.F. Macbr.

(Garapa) during soaking. **Revista Arvore**, Viçosa, MG, v. 26, n. 5, p. 593-601, 2002.

REIS, E. G. R. et al. Physiological quality of osmoprimed eggplant seeds. **Ciência e Agrotecnologia**, Lavras, v. 36, n. 5, p. 526-532, 2012.

ROSSETTO, C. A. V.; LIMA, T. M.; NAKAGAWA, J. Physiological quality and storage potential of tomato seeds submitted to osmotic conditioning. **Horticultura Brasileira**, Brasilia, v. 20, n. 4, p. 630-634, 2002.

SHIGEOKA, S. et al. Regulation and function of ascorbate peroxidase isoenzymes. **Journal of Experimental Botany**, Oxford, v. 53, n. 372, p. 13051319, 2002.

SUN, H. et al. Ascorbate-glutathione cycle of mitochondria in osmoprimed soybean cotyledons in response to imbibitional chilling injury. **Journal of Plant Physiology,** Minneapolis, v. 168, n. 3, p. 226-232, 2011.

TULLIO, M. C.; ARRIGONI, O. The ascorbic acid system in seeds: to protect and to serve. **Seed Science Research**, Wallingford, v.13, n. 4, p. 249-260, 2003.

YEMM, E. W.; COCKING, E. C.The determination of amino acid with ninhydrin. **Analyst**, London, v. 80, p. 209-213, 1955.

WEITIN, N. I.; TRELEASE, R. N.; EISING, R. Two temporally synthesized charge subunits interact to form the five isoforms of cotton seed (*Gossypium hirsutum)* catalase. **Biochemical Journal**, London, v. 269, n. 1, p. 233-238, 1990.

ARTICLE 2 INFLUENCE OF OSMOTIC CONDITIONING ON TOMATO SEED GERMINATION UNDER DIFFERENT ENVIRONMENTAL CONDITIONS

SUMMARY

Osmotic conditioning is a technique that can be used to increase the tolerance of seeds to possible stresses during germination and the initial development phase. The aim was to study the influence of the conditioning agents PEG, KNO_3 or both on the concentration of some carbohydrates and on tolerance to different environmental conditions. Santa Clara tomato seeds were conditioned as follows: 100%PEG 2 days; 50%PEG+50%KNO3 (v/v); 100% KNO3 2 days; 100% PEG 4 days; 50%PEG+50%KNO3 (v/v) 4 days; 100% KNO3 4 days; 100% PEG 6 days; 50%PEG+50%KNO3 (v/v) 6 days; 100% KNO3 6 days. After conditioning, the seeds were subjected to physiological analysis and analysis to quantify some carbohydrates. Conditioning in 100% KNO_3 for 4 days gave the highest IVG, %G and %P. Only conditioning in 100% PEG for 4 and 6 days did not increase the IVG at low temperature compared to the control. Under ideal conditions, regardless of conditioning, there was an increase in GVI. In addition, a lower concentration of total soluble sugars was observed in the conditioned seeds when compared to the control. Osmotic conditioning increased the germination rate of tomato seeds under ideal conditions, water stress and low temperature.

Keywords: Abiotic stress. Conditioning. Seed. Tomato.

1 INTRODUCTION

Currently, with population growth, there is an increasing need to seek substantial increases to meet the world's demand for food. However, the main limitations are concentrated in edaphoclimatic factors, especially water, temperature, among others, which lead seeds, seedlings and plants into situations of physiological stress. When present from the initial development of the seedling, they can reduce the speed index as well as the germination percentage of the seeds, which leads to unevenness and reduced *stand* and, consequently, irreparable losses in production.

Techniques have been studied to help seeds better overcome the crucial phase of germination and initial seedling development, increasing their tolerance to abiotic stresses. These include conditioning, based on the absorption of water by the seeds, but in a controlled manner, using compounds that reduce the potential of the solution. In this way, germination processes will only begin, but germination will not take place (HEYDECKER; GIBBINS, 1978). The most commonly used solutions are inert, such as polyethylene glycol and mannitol, due to their high molecular weight. Other compounds can be recommended, such as potassium nitrate and sodium chloride (NASCIMENTO, 2004).

Osmotic conditioning also provides biochemical benefits such as increased expression of LEAs (*late embryogenesis abundant*) proteins (SOEDA et al., 2005), increased activity of the antioxidant system (CHEN; ARORA, 2011; KIBINZA et al., 2011), as well as increasing the amount of adenosine triphosphate (ATP) and the ATP / ADP (adenosine diphosphate) ratio (CORBINEAU et al., 2000), stimulating cell division until development (G2 phase), (CASTRO et al., 2000), in addition to changes in reserve metabolism and cell homeostasis (FAROOQ et al., 2010). These benefits enable better physiological vigor, reflected in increases in germination speed, uniformity and seedling vigor, as well as expanding the range of possible environments for germination and initial growth (VARIER; VARI; DADLANI, 2010).

The possibility of germination and initial growth under adverse conditions suggests greater tolerance to abiotic stresses. In this context, Amini (2011), working with rye seeds, found significant results with increased tolerance to water deficit in *Capsicum annuum* L. Yadav, Kumari and Ahmed (2011) observed increased tolerance to cold and salt stress and, in tomatoes, Nawaz et al. (2012) found an increase in the vigor of tomato seeds under conditions of salt stress. In addition, abiotic stress is directly related to changes in carbohydrate metabolism. These have the function of stabilizing proteins and protein complexes, as well as maintaining membrane stability and the cell's osmotic potential lower than the external environment in order to maintain the water balance (BOHNERT; SHEN 1999; GUPTA; KAUR, 2005).

The aim was to study the influence of the conditioning agents PEG, KNO3 or both on the concentration of some carbohydrates and on tolerance to different environmental conditions.

2 MATERIAL AND METHODS

Tomato seeds (*Lycopersicon esculentum* Mill.) cv Santa Clara were conditioned in solutions of polyethylene glycol 6000 (PEG) and potassium nitrate (KNO_3) for 2, 4 and 6 days. The osmotic agent solutions (v/v) were: 100% PEG for 2, 4 and 6 days; 100% KNO_3 for 2, 4 and 6 days; and combinations of 50% PEG + 50% KNO_3 for 2, 4 and 6 days. All the solutions had a Ψs of - -1.1 MPa, equivalent to a concentration of 0.05M PEG and 0.22M KNO_3 . The seeds were incubated under constant light and aeration, at a temperature of 15 °C, with 15 mL of solution per gram of seed. Osmotic potential calculations were obtained using the Michel and Kaufmann equation (1973) for PEG6000 and the Van'Thoff equation (HILLEL, 1971) for KNO_3 . The control was not conditioned. After conditioning, the seeds were washed under running water to remove the osmoconditioning solution from the surface and dried in a room with a temperature of 25 °C and relative humidity of 50% for a period of twenty-four hours, when the seeds reached a humidity level of between 5-7%.

The germination rate was based on three replicates of 50 seeds placed on a paper substrate. They were moistened with distilled water or certain solutions depending on the stress, equivalent to 2.5 times the weight of the dry paper and kept at a temperature of 20-30 °C (ideal temperature) and a 12-hour photoperiod according to Brasil (2009).

Germination was counted daily to calculate the germination speed index (GVI) according to Maguire (1962). The final percentage of germination and normal seedlings was calculated after they had stabilized. The average dry mass was for the whole seedling, results presented in (mg), obtained from five repetitions of one seedling, 15 days after planting.

To assess germination at ideal temperature (IT), the seeds were placed in Petri dishes® on 3 sheets of *Germitest*® paper moistened with distilled water and kept in B.O.D., according to Brasil (2009). For germination at low temperature (TB), with the exception of the temperature, which was 15°C, the same conditions as for IT were maintained. However, for germination under salt (ES) and water (EH) stresses, a pre-test was first carried out with seeds without any treatment to determine at which potential there was a marked decrease in germination. This potential was used in the germination studies, in both cases using a potential of -0.6MPa. The paper was moistened with NaCl to simulate salt stress and PEG to simulate water stress, and then only distilled water was added. The other conditions for germination were the same as for TI.

The extracts for the analysis of starch, reducing sugars (RA), non-reducing sugars (NR) and total soluble sugars (AST) were obtained by macerating 200 mg of seeds in 5 mL of potassium phosphate buffer (100 mM; pH 7.0) and subjecting them to a water bath at 40^0 C for 30 minutes, centrifuged at 10,000 rpm for 10 minutes. After collecting the supernatant, the precipitate was resuspended in 5 mL of potassium phosphate buffer (100 mM; pH 7.0) and the supernatant was collected again and added to the first. The precipitate was used to quantify total soluble sugar and non-reducing sugar using the anthrone method according to Yemm and Willis (1954) and reducing sugar using the Miller method (1959).

The *pellet* was used to determine the starch content by resuspending it in 8 mL of potassium acetate buffer (200 mM; pH 4.8) and incubating it in a water bath at 40^0 C for 15 minutes. Then 2 mL of the enzyme amyloglucosidase (1 mg/mL in potassium acetate buffer) was added, and it was incubated again in a water bath at 40^0 C for 120 minutes, followed by centrifugation at 10000 rpm for 20 minutes (RICKARD; BEHN, 1987), and the content was determined using the anthrone method.

The experimental design of the trials was DIC (Completely Randomized Design), following the factorial scheme (3x3) +1, being: 3 osmotic solutions (100% PEG, 50%PEG+50%KNO3 and 100% KNO3); 3 incubation times (2, 4, 6 days) plus the control (unconditioned seeds). Statistical analysis was carried out using the *Sisvar* statistical program (FERREIRA, 1999) and the means were compared using the *Scott-Knott* test at a 5% probability of error ($P<0.05$).

3 Results and discussion

Seeds conditioned with 100%PEG, regardless of the treatment period, and 50%PEG+50%KNO3 4 days, had the lowest concentrations of reducing sugars (RA) (around 13%) compared to the control, while the rest did not differ from the control (Table 1). However, seeds conditioned in 100% PEG for 4 days, 100% KNO3 for 6 days and the control had the highest concentrations of non-reducing sugars (NR), while those conditioned in 100% PEG for 6 days had the lowest (around 25%) compared to the control (Table 1). In tomato seeds, Nawaz et al. (2011) observed higher amounts of AR, NR and AST sugars in seeds conditioned with KNO_3 compared to the control, attributing these results to increased alpha-amylase activity. However, in this study, this type of relationship was not clear. The highest AST concentration was observed in the control at over 17% in relation to the others (Table 1). The reduction in AST in the conditioned seeds may be due to the mobilization of these carbohydrates for metabolic demand during the germination process (CORTE et al., 2006), since the lowest germination speed index (GVI) was observed for the control, as can be seen (Table 5). As for starch, an increase of approximately 35% was observed in seeds conditioned with 50%PEG+50%KNO3 for 6 days compared to the control (Table 1). Similar results were found by Magalhâes, Borges and Berger (2010) working with *Schizolobium parahyba* (Vell.) seeds and Corte et al. (2006) working with *Caesalpinia peltophoroides* (Benth.) seeds in which both observed a reduction in AST in the conditioned seeds. However, Magalhães, Borges and Berger (2010) recorded a reduction in starch content throughout germination, while Corte et al. (2006) observed little variation in the content of this compound throughout germination.

Table 1 Quantidate of reducing sugar (RA), non-reducing sugar (NR), sugar total soluble (AST), and starch in different Conditionings or without conditioning, result in gram/gram of seed dry matter - UFLA - 2013.

Treatments	AR	NR	AST	STARCH
Witness	0.039 a	0.50 a	0.092 a	0.078 b
100%PEG/4 d	0.033 b	0.49 a	0.068 b	0.082 b
50%PEG+50%KNO3/4 d	0.030 b	0.43 b	0.077 b	0.086 b
100%KNO3/4 d	0.039 a	0.43 b	0.075 b	0.084 b
100%PEG/6 d	0.034 b	0.37 c	0.072 b	0.085 b
50%PEG+50%KNO3/6 d	0.042 a	0.44 b	0.074 b	0.102 a
100%KNO3/6 d	0.043 a	0.46 a	0.069 b	0.091 b
hp %	16,78	7,98	13,04	8,66

Averages followed by the same letters in the rows and columns do not differ according to the *Scott-Knott* test at 5% probability.

In the physiological variables evaluated for water stress, conditioning with 100% KNO3 for 4 days increased IVG by 62%, %G by 21% and %P by 25% compared to the control. In addition, 100%PEG 4-day conditioning was the most advantageous (Table 2). In general, conditioning increased physiological vigor. Amini (2011) working with rye seeds under water stress, conditioned in PEG, observed an increase in the length of the aerial part and root as well as in the speed of germination. According to Amini (2011), the benefits came from an increase in the activity of invertase and amylase, which resulted in a higher concentration of cellular solutes, thus decreasing the internal cellular potential, which maintained the water balance.

Table 2 Physiological analysis of conditioned tomato seeds subjected to water stress at -0.6 MPa - UFLA - 2013.

Treatments	IVG	%G	%P	M. M.
Witness	7.09 d	71 b	60 b	3,52 a
100%PEG/4 d	8.52 d	77 b	76 b	4,37 a
50%PEG+50%KNO3/4 d	14.56 c	83 b	81 a	5,12 a
100%KNO3/4 d	19.18 a	89 a	79 a	4,57 a
100%PEG/6 d	16.03 b	81 b	79 a	4,57 a
50%PEG+50%KNO3/6 d	13.72 c	88 b	80 a	5,30 a
100%KNO3/6 d	13.48 c	83 b	77 a	4,64 a
cv%	9,19	4,09	12, 97	16,57

Means followed by the same letters in the columns do not differ by the Scott-Knott test at 5% probability. germination speed index (IVG), seedling formation speed index (IVP); germination percentage (%G), normal seedling percentage (%P), average seedling mass (M. M.) in mg.

In relation to salt stress, there was only a difference in the mass measurements, where it was observed that the 50%PEG+50%KNO3 4-day, 100%PEG 6-day and 100%KNO3 6-day conditionings showed averages around 18% lower than the control (Table 3). Nawaz et al. (2012) worked with tomato seeds conditioned in sand and sodium chloride, under conditions of salt stress, and observed an increase in the IVG, %G and vigor of the seedlings, justifying this response to the better functioning of the antioxidant system. However, the use of sodium chloride by Nawaz et al. (2012) in conditioning possibly induced different metabolic responses compared to the present study, which allowed for greater tolerance to salt stress.

Table 3 Physiological analysis of conditioned tomato seeds subjected to salt stress at -0.6 MPa - UFLA - 2013.

Treatments	IVG	%G	%P	M. M.
Witness	6.87 a	78 a	67 a	6,29 a
100%PEG/4 d	7.00 a	75 a	73 a	6,34 a
50%PEG+50%KNO3/4 d	7.95 a	76 a	74 a	4,56 b

100%KNO3/4 d	8.97 a	74 a	68 a	5,58 a
100%PEG/6 d	10.04 a	88 a	84 a	5,21 b
50%PEG+50%KNO3/6 d	7.57 a	80 a	78 a	6,17 a
100%KNO3/6 d	8.61 a	70 a	64 a	4,14 b
cv	18,13	10,62	13,22	12,45

Averages followed by the same letters in the columns do not differ by the *Scott-Knott* test at 5% probability. germination speed index (IVG), germination percentage (%G), normal seedling percentage (%P), average seedling mass (M. M.) in mg.

With regard to the physiological analyses observed for low temperature stress, the lowest IVG values were observed in the control and when conditioned in PEG for 4 and 6 days. However, the other conditionings resulted in increases in GVI of at least 20% compared to the control (Table 4). Similar results were found by Nascimento (2005) working with tomato seeds, where he observed higher values of GVI in seeds conditioned with KNO_3, when compared to the control. However, there was an upward trend in all the analyses carried out for the conditioned seeds compared to the control. Amooaghaie, Nikzad and Shareghi (2010) worked with tomatoes conditioned with PEG under low temperature conditions and attributed the increase to the speed of emergence and the average emergence rate. They also reported changes in the amount of soluble protein and phosphorus, which resulted in better membrane stability and integrity.

Table 4 Physiological analysis of conditioned tomato seeds subjected to low temperature stress at 15 °C - UFLA - 2013.

Treatments	IVG	%G	%P	M. M.
Witness	3.94 b	81 a	78 a	3,39 a
100%PEG/4 d	4.84 b	82 a	80 a	5,64 a
50%PEG+50%KNO3/4 d	5.81 a	88 a	80 a	5,09 a
100%KNO3/4 d	4.97 a	89 a	86 a	3,85 a
100%PEG/6 d	4.34 b	81 a	76 a	3,12 a
50%PEG+50%KNO3/6 d	5.13 a	83 a	80 a	4,08 a

100%KNO3/6 d	5.62 a	88 a	85 a	3,49 a
cv	8,95	7,44	8,94	26,93

Averages followed by the same letters in the columns do not differ by the Scott-Knott test at 5% probability. germination velocity index (IVG), germination percentage (%G), normal seedling percentage (%P), average seedling mass (M. M.) in mg.

Under ideal conditions, the control had the lowest GVI. In addition, the 100%KNO3 6-day and 50%PEG+50%KNO3 4-day conditions were the ones that gave the highest GVI, around 43% higher than the control (Table 5). There were no differences in the other analyses. Farooq et al. (2005) worked with tomato seeds conditioned in PEG and KNO3 and found similar results, where the GVI was higher for those conditioned in 100%KNO_3 , followed by 100%PEG and, lastly, the control. However, Farooq et al. (2005) found an increase in growth, but in the present study, the height averages did not differ from the control. However, the difference may be due to the use of poorer quality seeds, since the average germination percentage in the study by Farooq et al. (2005) was 50% lower than in the present study. According to Varier, Vari and Dadlani (2010), the effects of conditioning are more pronounced in seed batches of poorer quality.

Table 5 Physiological analysis of conditioned tomato seeds submitted to ideal conditions - UFLA - 2013.

Treatments	IVG	%G	%P	M. M.
Witness	12,07 c	87 a	85 a	3,41 a
100%PEG/4 d	17.00 b	86 a	84 a	4,91 a
50%PEG+50%KNO3/4 d	21.70 a	87 a	85 a	4,52 a
100%KNO3/4 d	17.86 b	87 a	81 a	4,53 a
100%PEG/6 d	15.22 b	81 a	78 a	4,09 a
50%PEG+50%KNO3/6 d	16.40 b	84 a	80 a	4,26 a
100%KNO3/6 d	21.34 a	86 a	84 a	5,11 a
cv	8,72	5,59	6,32	15,70

Averages followed by the same letters in the columns do not differ by the Scott-Knott test at 5%

probability. germination velocity index (IVG), germination percentage (%G), normal seedling percentage (%P), average seedling mass (M. M.) in mg.

In general, conditioning led to an increase in IVG under ideal conditions and for water and low temperature stress. Conditioning was ineffective only under salt stress. There was no clear relationship between carbohydrates and physiological responses to stress. It was only under ideal conditions that lower AST concentrations were observed in seeds with higher GVI. The osmotic conditioning technique is very promising in terms of increasing the tolerance of initial seed development, but more studies are needed to find the best incubation medium for each species and each type of stress.

4 CONCLUSION

Conditioning 100%KNO3 for 4 days is the most suitable for increasing IVG, %G and %P under water stress. Conditioning was not efficient under salt stress. Only conditioning in 100% PEG for 4 and 6 days was not effective in increasing IVG at low temperatures. Under ideal conditions, regardless of conditioning, there was an increase in GVI. Finally, there was a decrease in the concentration of total soluble sugar in conditioned tomato seeds.

REFERENCES

AMINI, R. Effects of osmopriming on drought stress tolerance of perennial rye (Secale *montanum* Guss.) during germination. **Journal of Food Agriculture and Environment**, Helsinki, v. 9, p. 305-308, 2011.

AMOOAGHAIE, R.; NIKZAD, K.; SHAREGHI, B. The effect of priming on emergence and biochemical changes of tomato seeds under suboptimal temperatures. **Seed Science and Technology**, Zurich, v. 38, p. 508-512, 2010.

BOHNERT, H. J.; SHEN, B. Transformation and compatible solutes. **Scientia Horticulturae**, Amsterdam, v. 78, p. 237-260, 1999.

BRAZIL. Ministry of Agriculture, Livestock and Supply. **Rules for seed analysis.** Brasilia, 2009. 395 p.

CASTRO, R. D. et al. Cell division and subsequent radicle protrusion in tomato seeds are inhibited by osmotic stress but DNA synthesis and formation of microtubular cytoskeleton are not. **Plant Physiology**, Minneapolis, v.122, p. 327-336, 2000.

CHEN, K.; ARORA, R. Dynamics of the antioxidant system during seed osmopriming, post-priming germination, and seedling establishment in Spinach (*Spinacia oleracea*). **Plant science**, Limerick, v. 180, p. 212-220, 2011.

CORBINEAU, F. et al. Improvement of tomato seed germination by osmopriming as related to energy metabolism. In: INTERNATIONAL WORKSHOP ON SEEDS, 6., 1999, Merida. **Proceedings...** Cambridge: CABI, 2000, p. 467-474.

CORTE, V. B. et al. Mobilization of reserves during seed germination and seedling growth of *Caesalpinia peltophoroides* Benth.

(Leguminosae-Caesalpinoideae). **Revista Arvore**, Viçosa, MG, v. 30, p. 941949, 2006.

FAROOQ, M. et al. Thermal hardening: a new seed vigor enhancement tool in rice. **Journal of Integrative Plant Biology**, Chichester, v. 47, p. 187-93, 2005.

FAROOQ, M. et al. Changes in nutrient-homeostasis and reserves metabolism during rice seed priming: consequences for seedling emergence and growth.

Agricultural Sciences in China, Beijing, v. 9, p. 191-198, 2010

FERREIRA, D. F. **Sisvar**: statistical analysis system. Lavras: UFLA, 1999.

GUPTA, A. K.; KAUR, N. Sugar signaling and gene expression in relation to carbohydrate metabolism under abiotic stresses in plants. **Journal of Biosciences**, Bangalore, v. 30, p. 761-776, 2005.

HEYDECKER, W.; GIBBINS, B. M. The priming of seeds. **Acta Horticulturae**, The Hague, v. 83, p. 213-223, 1978.

HILLEL, D. The state of water in the soil. In:. **Soil and water:** physical principles and processes. New York: Academic, 1971. p. 49-77.

KIBINZA, S. et al. Catalase is a key enzyme in seed recovery from ageing during priming. **Plant science**, Limerick, v. 181, p. 309-315, 2011.

MAGALHAES, S. R.; BORGES, E. E. L.; BERGER, A. P. A. Mobilization of reserves in the embryonic axis and cotyledons of *Schizolobium parahyba* (Vell.) S. F. Blake seeds during germination. **Ciência Florestal**, Santa Maria, v. 20, p. 589-595, 2010.

MAGUIRE, J. D. Speed of germination-aid in relation evaluation for seedling emergence vigor. **Crop Science**, Madison, v. 2, p. 176-177, 1962.

MILLER, G. Use of dinitrosalicylic acid reagent for determination of reducing sugars. **Analytical Chemistry**, Washington, v. 31, p. 426-428, 1959.

NASCIMENTO, W. M. **Osmotic conditioning of vegetable seeds**. Brasilia: Embrapa Hortaliças, 2004. 12 p. (Technical Circular, 33).

NASCIMENTO, W. M. Osmotic conditioning of vegetable seeds for germination under low temperature conditions. **Horticultura Brasileira,** Brasilia, v. 23, p. 211-214, 2005.

NAWAZ, A. et al. Effect of halopriming on germination and seedling vigor of tomato. **African Journal of Agricultural Research**, Ekapoma, v. 6, p. 35513559, 2011.

NAWAZ, A. et al. Induction of salt tolerance in tomato (*Lycopersicon esculentum* Mill .) seeds through sand priming. **Australian Journal of Crop Science**, Lismore, v. 6, p. 1199-1203, 2012.

RICKARD, J. E.; BEHN, K. R. Evaluation of acid and enzyme hydrolitic methods for determination of cassava starch. **Journal of the Science of Food and Agriculture**, London, v. 41, n. 4, p. 373-379, 1987.

SOEDA, Y. et al. Gene expression programs during *Brassica oleracea* seed maturation , osmopriming , and germination are indicators of progression of the germination process and the stress tolerance level. **Plant Physiology**, Minneapolis, v. 137, p. 354-368, 2005.

VARIER, A.; VARI, A. K.; DADLANI, M. The subcellular basis of seed priming. **Current science**, Bangalore, v. 99, p. 450-456, 2010.

YADAV, V. P.; KUMARI, M.; AHMED, Z. Seed priming mediated germination improvement and tolerance to subsequent exposure to cold and salt stress in Capsicum. **Research Journal of Seed Science**, Cambridge, v. 3, p. 125-136, 2011.

YEMM, E. W.; WILLIS, A. J. The estimation of carbohydrates in plant extracts by anthrone. **Biochemical Journal,** London, v. 57, p. 508-515, 1954.

Printed by Books on Demand GmbH, Norderstedt / Germany